WILD STEM

STEM in the GARDEN

Megan Borgert-Spaniol

Checkerboard Library

An Imprint of Abdo Publishing
abdobooks.com

abdobooks.com

Published by Abdo Publishing, a division of ABDO, PO Box 398166, Minneapolis, Minnesota 55439.

Printed in the United States of America, North Mankato, Minnesota
102023
012024

THIS BOOK CONTAINS RECYCLED MATERIALS

Design: Denise Hamernik, Mighty Media, Inc.
Production: Mighty Media, Inc.
Editor: Anna Anderhagen
Cover Photographs: POP-THAILAND/Shutterstock Images
Interior Photographs: anut21ng Stock/Shutterstock Images, p. 29; Ashlee Ruggels/AP Images, p. 7; D. Ross Cameron/AP Images, p. 27; Dmytro Zinkevych/Shutterstock Images, p. 15; Em Campos/Shutterstock Images, p. 21; Kikujiarm/Shutterstock Images, p. 13; Mighty Media, Inc., pp. 16–17, 22–23; 1000 Words/Shutterstock Images, p.19; Stephen Davies/Adobe Stock, p. 11; tawanroong/Shutterstock Images, p. 25; Thomas Söllner fotodesign/Adobe Stock, p. 9; Кирилл Рыжов/Adobe Stock, p. 5
Design Elements: Lorthois Yuliya/Shutterstock Images; supanut/Adobe Stock

Library of Congress Control Number: 2023939341

Publisher's Cataloging-in-Publication Data
Names: Borgert-Spaniol, Megan, author.
Title: STEM in the garden / by Megan Borgert-Spaniol
Description: Minneapolis, Minnesota : Abdo Publishing, 2024 | Series: Wild STEM | Includes online resources and index.
Identifiers: ISBN 9781098291969 (lib. bdg.) | ISBN 9781098278861 (ebook)
Subjects: LCSH: Science--Study and teaching--Juvenile literature. | Technology--Study and teaching--Juvenile literature. | Engineering--Study and teaching--Juvenile literature. | Mathematics--Study and teaching--Juvenile literature. | Gardening--Juvenile literature. | Plants--Juvenile literature.
Classification: DDC 577.0--dc23

CONTENTS

WILD GARDEN

Do you love the smell of sunbaked soil? Do you dream of spending days surrounded by flowers and other plants? If so, you might enjoy a career that takes you to the garden!

Many people work in gardens. Landscaping professionals design and construct garden spaces. Gardeners and groundskeepers maintain landscaped areas. Horticulturists are experts in plant care, while arborists **specialize** in tree care. Nursery workers grow plants outdoors and in greenhouses.

What kind of job would you like to have? There's only one way to find out. Pull on your gardening gloves and grab a rake. It's time to explore science, technology, engineering, and math (STEM) in the garden!

STEM
Science
Technology
Engineering
Math

Gardens remove air pollution, provide fresh food, and create homes for wildlife.

HORTICULTURISTS

Gardens are sources of food and beauty. The science of growing and maintaining gardens is called horticulture. People who **specialize** in this practice are called horticulturists.

Horticulturists are experts in all areas of plant care, from soil health to fertilization. They identify harmful insects and diseases. They are skilled in plant propagation. Plant propagation is growing new plants from existing plant parts, such as stems or leaves.

Many horticulturists create new plant varieties. They may produce a new type of tomato that is more flavorful or create a flower that is more disease **resistant**. These scientists are valuable sources of knowledge for anyone who wants a thriving garden!

Horticulturists at the Royal Botanic Gardens in England examine the leaves of a new species of giant water lily.

ARBORISTS

Trees add beauty and structure to gardens. They also provide **habitats** for insects, birds, and other wildlife. Sometimes, trees are threatened by damage or disease. In these cases, trees and gardens benefit from the knowledge and skill of arborists.

Arborists are arboriculture experts. Arboriculture is the **cultivation** of trees, shrubs, and other woody plants. Arborists advise people on what trees to plant and where to plant them. They inspect existing trees, **diagnose** diseases, and control pest problems. Arborists **prune** and fertilize trees, too.

Arborists do everything they can to keep trees alive and healthy. They spray trees with chemicals to get rid of harmful insects or diseases. Arborists use cables and rods to support weak branches, making them stronger.

An arborist uses a chain saw to prune part of a broken limb.

Unfortunately, not all trees can be saved. Sometimes trees spread disease or cause property damage. In these cases, arborists know how to safely cut down trees without harming surrounding trees and structures.

Caring for trees requires many different tools. Arborists use ropes, harnesses, and other climbing gear to climb trees safely. They use saws and tree loppers to **prune** or remove trees. Arborists wear helmets, goggles, and other protective gear when doing this work. Their **dedication** to tree care keeps yards and gardens strong!

Being an arborist requires physical fitness and being comfortable with heights.

NURSERY WORKERS

Starting a garden takes time and effort. Some gardeners grow their plants from seeds. But many gardeners purchase plants that have already begun to grow at nurseries.

Nurseries are businesses that grow plants, including flowers, vegetables, shrubs, and trees. Nurseries sell their young plants in large quantities to landscaping companies or retail garden centers. Some nurseries also sell their plants directly to individual customers.

Nursery workers grow plants outdoors and indoors. Trees, shrubs, and **perennial** flowers grow outdoors. Most **annual** flowers and garden vegetables are too sensitive to grow outdoors year-round. These plants get started in greenhouses.

Nursery workers organize young plants by type and season. They ensure the plants get the correct amount of sunlight in the greenhouse.

Nursery workers also prepare the soil for planting seeds. They water, **prune**, and fertilize growing plants. Sometimes they spray plants with **pesticides** and other chemicals to kill harmful insects, weeds, and diseases. When seedlings get too big for their containers, workers **transplant** them into larger containers.

Nursery workers use **trowels**, pruning shears, and other hand tools for their work. Larger operations use bigger equipment, such as soil mixers and potting machines. Nursery workers use hoses, pumps, and sprinklers for plant **irrigation**. Thanks to nursery workers, gardens are filled with healthy, beautiful plants!

Nursery workers check the temperature for the sensitive vegetables grown in a greenhouse.

GET WILD!

GREENHOUSE GROWING

Nursery workers control the temperature and humidity inside greenhouses. This experiment shows how controlling these conditions allows plants to thrive.

WHAT YOU NEED:

- 2 clear plastic bottles with caps
- craft knife
- potting soil
- beans or other seeds
- water
- tape
- tray

WHAT YOU DO:

1. Have an adult help you use a craft knife to poke some holes in the bottom of each bottle. Then have them cut the bottles in half to separate tops from bottoms.

2. Fill both bottle bottoms halfway with potting soil. Set the bottles on the tray.

3. Place three seeds in each bottle. Cover the seeds with ¼ inch (0.6 cm) of soil.

4. Water the soil in both bottles so it is damp. Attach the top portion of one of the bottles back onto the bottom portion using tape. This covered bottle represents a greenhouse.

5. Place the covered and uncovered bottles on the tray next to a sunny window. Check the bottles every day.

6. Observe what happens. Do sprouts appear in one bottle before the other? Do the sprouts in one bottle grow faster than those in the other? What can you conclude about growing plants in a greenhouse compared to outdoors?

LANDSCAPERS

Gardens range from small and simple to large and grand. Gardens grow in private yards and public parks. Some landowners seek help from professionals in building their gardens. These professionals are called landscapers.

Landscape designers are one type of landscaper. They plan and create gardens for small residential spaces, such as backyards. They design these spaces to meet their clients' goals and budgets.

Landscape design involves selecting and placing flowers, trees, and other plants. These living elements of a garden are called softscape. Some landscape designers also work with nonliving elements, such as pathways, patios, and firepits. Nonliving elements of a garden are called hardscape.

A landscape designer created this backyard garden by using softscape elements to add color and privacy.

Another type of landscaper is a landscape architect. Landscape architects are trained to plan larger gardens, such as those at parks, college campuses, and resorts. They may use both softscape and hardscape in their designs.

Landscapers use many tools to plan garden spaces. They use laser **levels** to learn the slope of the land. They also create 3D models of their designs using computer **software**.

Landscape designers and architects work with landscaping crews to carry out their plans. These crews use hand tools and powerful machines for digging, raking, and leveling. Landscapers take great care to create gardens that exist in harmony with people.

Landscape architects helped lead the design of the indoor gardens at Jewel Changi Airport in Singapore.

GET WILD!

POND IN A POT

Landscaping professionals may include ponds in their designs. Build your own pond in a pot!

WHAT YOU NEED:

- ✔ large, nonporous washtub or plant pot
- ✔ bricks or large stones
- ✔ small rocks, bark, or sticks
- ✔ 2 aquatic plants (such as marsh marigold or cattails)
- ✔ collected rainwater
- ✔ 2 floating plants (such as duckweed or frogbit)

WHAT YOU DO:

1. Place the tub outdoors in a spot with shade. Place a few bricks or large stones in the bottom of the tub.

2. Stack bricks or stones to the rim of the tub. The bricks or stones will give small creatures a way out of the tub. Stack rocks, bark, or sticks on the outside of the tub too. It will be easier for critters to crawl in that way.

3. Arrange the aquatic plants in the tub. Place small rocks on top of the roots of the plants to hold them in place.

4. Fill the tub with rainwater so all the plant roots are submerged.

5. Fill in space around the aquatic plants with the floating plants. Observe what kinds of critters visit!

GARDENERS & GROUNDSKEEPERS

While landscapers plan and create garden spaces, gardeners and groundskeepers maintain them. Gardeners and groundskeepers take care of the plants and other features. They ensure university grounds, public gardens, golf courses, and other spaces look their best.

Gardeners plant **annual** and **perennial** plants. They must remove weeds and ensure plants are fully watered and fertilized. Gardeners also mow grass and trim bushes.

Groundskeepers perform upkeep beyond gardening duties. They make sure light fixtures, sprinklers, and fountains are working. Groundskeepers may also empty garbage bins or clear snow from sidewalks.

A gardener records the growth of plants and looks for disease in a sunflower garden.

Groundskeepers and gardeners may work at parks, hotels, or sports fields. Others work for botanical gardens, **arboretums**, or **cemeteries**. Some zoos, hospitals, and universities that have **extensive** grounds also employ gardeners and groundskeepers.

Gardeners and groundskeepers use the same tools for gardens and lawn care. Shovels, **trowels**, and rakes help prepare the soil for planting. Lawn **aerators** poke holes in the ground so water and nutrients can reach the grass roots. Leaf blowers remove leaves from gardens and paths in the fall. Gardeners and groundskeepers work hard to maintain stunning outdoor spaces!

Groundskeepers get the baseball diamond ready before a game between the Chicago White Sox and San Francisco Giants.

A CAREER IN THE GARDEN

Do you have a passion for STEM in the garden? There are many career paths to choose from. Each path leads to a wide range of opportunities.

A horticulturist might work for a nursery or retail garden center. A groundskeeper might work at a theme park or resort. An arborist might work for a university or city government. No matter where they work, these professionals share a love of our planet's gardens!

**Maybe one day, the garden will be your office!
What will you discover?**

GLOSSARY

aerator—a tool that supplies something with air.

annual—living for only one growing season.

arboretum—a place where trees, shrubs, and woody plants are grown for scientific and educational purposes.

cemetery—a burial ground.

cultivation—the act or art of fostering the growth of a plant or tree.

dedication—a commitment to a goal or a way of life.

diagnose—to recognize something, such as a disease, by signs, symptoms, or tests.

extensive—covering a large area.

habitat—a place where a living thing is naturally found.

humidity—the amount of moisture in the air.

irrigation—the watering of plants by artificial means to foster plant growth.

level—an instrument that shows whether a surface is flat and even.

perennial—living for more than one growing season.

pesticide—a substance used to destroy pests.

prune—to cut away or remove what is extra or unwanted.

resistant—not harmed or affected by something.

software—the written programs used to operate a computer.

specialize—to concentrate one's efforts in a specific area.

transplant—to move something from one place to another.

trowel—a gardening tool similar to a small shovel.

ONLINE RESOURCES

Booklinks
NONFICTION NETWORK
FREE! ONLINE NONFICTION RESOURCES

To learn more about STEM in the garden, visit **abdobooklinks.com**. These links are routinely monitored and updated to provide the most current information available.

INDEX